Harikumar Rajaguru
Sunil Kumar Prabhakar

KNN Classifier and K-Means Clustering for Robust Classification of Epilepsy from EEG Signals

A Detailed Analysis

Anchor Academic
Publishing

Rajaguru, Harikumar, Prabhakar, Sunil Kumar: KNN Classifier and K-Means Clustering for Robust Classification of Epilepsy from EEG Signals. A Detailed Analysis, Hamburg, Anchor Academic Publishing 2017

Buch-ISBN: 978-3-96067-140-4
PDF-eBook-ISBN: 978-3-96067-640-9
Druck/Herstellung: Anchor Academic Publishing, Hamburg, 2017

Bibliografische Information der Deutschen Nationalbibliothek:
Die Deutsche Nationalbibliothek verzeichnet diese Publikation in der Deutschen Nationalbibliografie; detaillierte bibliografische Daten sind im Internet über http://dnb.d-nb.de abrufbar.

Bibliographical Information of the German National Library:
The German National Library lists this publication in the German National Bibliography. Detailed bibliographic data can be found at: http://dnb.d-nb.de

All rights reserved. This publication may not be reproduced, stored in a retrieval system or transmitted, in any form or by any means, electronic, mechanical, photocopying, recording or otherwise, without the prior permission of the publishers.

Das Werk einschließlich aller seiner Teile ist urheberrechtlich geschützt. Jede Verwertung außerhalb der Grenzen des Urheberrechtsgesetzes ist ohne Zustimmung des Verlages unzulässig und strafbar. Dies gilt insbesondere für Vervielfältigungen, Übersetzungen, Mikroverfilmungen und die Einspeicherung und Bearbeitung in elektronischen Systemen.

Die Wiedergabe von Gebrauchsnamen, Handelsnamen, Warenbezeichnungen usw. in diesem Werk berechtigt auch ohne besondere Kennzeichnung nicht zu der Annahme, dass solche Namen im Sinne der Warenzeichen- und Markenschutz-Gesetzgebung als frei zu betrachten wären und daher von jedermann benutzt werden dürften.

Die Informationen in diesem Werk wurden mit Sorgfalt erarbeitet. Dennoch können Fehler nicht vollständig ausgeschlossen werden und die Diplomica Verlag GmbH, die Autoren oder Übersetzer übernehmen keine juristische Verantwortung oder irgendeine Haftung für evtl. verbliebene fehlerhafte Angaben und deren Folgen.

Alle Rechte vorbehalten

© Anchor Academic Publishing, Imprint der Diplomica Verlag GmbH
Hermannstal 119k, 22119 Hamburg
http://www.diplomica-verlag.de, Hamburg 2017
Printed in Germany

ABSTRACT

Epilepsy is a chronic disorder, the hallmark of which is recurrent, unprovoked seizures. Many people with epilepsy have more than one type of seizures and may have other symptoms of neurological problems as well. Epilepsy is caused due to sudden recurrent firing of the neurons in the brain. The symptoms are convulsions, dizziness and confusion. One out of every hundred person experience seizure at sometime in their lives. It may be confused with other events like strokes or migraines. Unfortunately, the occurrence of an epileptic seizure seems unpredictable and its process is very little understood. In India number of persons are suffering from Epilepsy are increasing every year. The complexity involved in the diagnosis and therapy is to be cost effective in nature. In this project, we applied an algorithm which is used for classification of risk level of epilepsy in epileptic patients from Electroencephalogram (EEG) signals. Dimensionality reduction is done on the EEG dataset by applying Power Spectral density. The KNN Classifier and K-Means clustering is implemented on these spectral values to epilepsy risk level detection. The Performance Index (PI) and Quality Value (QV) are calculated for the above methods. A group of twenty patients with known epilepsy findings are used in this study. High PI such as 98.5% was obtained at QV's of 22.37, for K-Means Clustering when compared to the value of 18.02 through KNN Classifier respectively.

TABLE OF CONTENTS

ABSTRACT 1
TABLE OF CONTENTS 2
LIST OF TABLES 4
LIST OF FIGURES 5
LIST OF SYMBOLS 6
LIST OF ABBREVIATIONS 7

CHAPTER 1 INTRODUCTION 9
   1.1   FUNDAMENTALS OF EEG 10
   1.2   EPILEPSY TYPES AND SYMPTOMS 12
      1.2.1   Generalized seizures 12
      1.2.2   Partial or focal seizures 13
      1.2.3   Absence or petit mal seizures 13
   1.3   MATHEMATICAL APPROACH IN MEDICAL DIAGNOSIS 14
   1.4   EEG SIGNALS FOR EPILEPSY DETECTION 15
   1.5   ORGANIZATION OF THE PROJECT REPORT 18

CHAPTER 2 MATERIALS AND METHODS 20
   2.1   ACQUISITION OF EEG DATA 21
   2.2   DIMENSIONALITY REDUCTION 22
      2.2.1   POWER SPECTRAL DENSITY 25
      2.2.2   PERIODOGRAM 26
      2.2.3   WINDOW FUNCTION 27
      2.2.4   RECTANGULAR WINDOW 29

- CHAPTER 3 KNN CLASSIFIER 31
  - 3.1   OVERVIEW OF *K*-NEAREST NEIGHBOR ALGORITHM 31
  - 3.2   PARAMETER *K* 33
  - 3.3   DISTANCE METRICS 34
    - 3.3.1   Euclidean Distance Metric (Eu) 34
    - 3.3.2   City Block Distance Metric (Cb) 34
    - 3.3.3   Correlation distance metric (CO) 35
  - 3.4   KNN CLASSIFIER ALGORITHM 36

- CHAPTER 4 K- MEANS CLUSTERING 39
  - 4.1   INTRODUCTION TO K-MEANS CLUSTERING 39
  - 4.2   ALGORITHMIC STEPS FOR K-MEANS CLUSTERING 40
  - 4.3   DECIDING THE NUMBER OF CLUSTERS 41

- CHAPTER 5 RESULT AND DISCUSSION 44
  - 5.1   PERFORMANCE INDEX 44
  - 5.2   QUALITY VALUE 45

- CHAPTER 6 CONCLUSION AND FUTURE EXPANSIONS 48

- REFERENCES 49

LIST OF TABLES

Table 2.1 Target Values for Groups 23
Table 2.2 Features of various windowing techniques.. 28
Table 3.1 Performance parameters of KNN Classifier 37
Table 4.1 Performance parameters of K-Means clustering 42
Table 5.1 Performance Comparisons of KNN Classifier and K-Means Clustering 46

LIST OF FIGURES

Figure 1.1 EEG Waveform 11
Figure 2.1 EEG Recording by 10-20 system 20
Figure 2.2 Epileptic EEG Signal Waveform of Patient 21
Figure 2.3 Sample 2-second epoch 22
Figure 2.4 Flow diagram of the Epilepsy Risk Level Classification System 24
Figure 2.5 Power spectral density of a signal 26
Figure 3.1 KNN Classifier 33
Figure 4.1 K-Means clustering working 40
Figure 5.1 Sensitivity and Specificity measures of KNN classifier and
 K- means clustering 46
Figure 5.2 Quality factor and Time delay measures of KNN classifier and
 K- means clustering 47
Figure 5.3 Average detection and Quality factor measures of KNN classifier and
 K- means clustering 47

LIST OF SYMBOLS

α	Alpha
σ	Sigma
δ	Gamma
θ	Theta
ω	Omega
K	Kilo
Ω	Ohm

LIST OF ABBREVIATIONS

EEG	Encephalographer
EEGer	Electroencephalogram
ECOG	Electrocorticogram
Hz	Hertz
GA	Genetic Algorithm
NDS	Nonlinear Dynamic System
PSD	Power Spectral Density
KNN	K- Nearest Neighborhood
PI	Performance Index
PC	Perfect Classification
Qv	Quality value
ROC	Receiver Operating Characteristics

CHAPTER 1
INTRODUCTION

Epilepsy, from which approximate 1% of the people in the world suffer, is a group of brain disorders characterized by the recurrent paroxysmal electrical discharges of the cerebral cortex, that result in irregular disturbances of the brain functions, which are associated with the significant changes of the EEG signal Electroencephalograms (EEGs) are recordings of the electrical potentials produced by the brain[1]. Analysis of EEG activity has been achieved principally in clinical settings to identify pathologies and epilepsies since Hans Berger's recording of rhythmic electrical activity from the human scalp. In the past, interpretation of the EEG was limited to visual inspection by a neurophysiologist, an individual trained to qualitatively make a distinction between normal EEG activity and abnormalities contained within EEG records. It is known that biological neurons can be modeled by a set of nonlinear differential equations. The minimal embedding dimension gives the upper number of nonlinear dynamic system (NDS) freedom degrees and the minimal number of differential equations demanded for mathematical modeling of NDS. Therefore, the change of the structure of brain NDS during seizure can be shown by the change of embedding dimension of EEG signals if the human brain is considered as a nonlinear dynamic system. A common form of recording used for this purpose is an ambulatory recording that contains EEG data for a very long duration of even up to one week. It involves an expert's efforts in analyzing the entire length of the EEG recordings to detect traces of epilepsy [2]. Because seizures, in general, occur frequently and unpredictably, automatic detection of seizures during long term EEG monitoring sessions is highly useful and needed.

Electroencephalography (EEG) is an important clinical tool, monitoring, diagnosing and managing neurological disorders related to epilepsy. In comparison with other methods such as Electrocorticogram (ECOG), EEG is a clean and safe technique for monitoring the brain activity [11]. In spite of available dietary, drug and surgical treatment options, currently nearly one out of three epilepsy patients cannot be treated. They are completely subject to the sudden and unforeseen seizures which have a great effect on their daily life, with temporary impairments of perception, speech, motor control, memory and/or consciousness [3]. Many new therapies are being investigated and among them the most promising are implantable devices that deliver direct electrical stimulation to affected areas of the brain. These treatments will greatly depend on robust algorithms for seizure detection to perform effectively. Because the onset of the seizures cannot be predicted in a short period, a continuous re-cording of the EEG is required to detect epilepsy. How-ever, analysis by visual inspection of long recordings of EEG, in order to find traces of epilepsy, is tedious, time- consuming and high-cost. Therefore, automated detection of epilepsy has been a goal of many researchers for a long time. With the advent of technology, the digital EEG data can be input to an automated seizure detection system, allowing physicians to treat more patients in a given time because the time taken to review the EEG data is greatly reduced by automation.

1.1 FUNDAMENTALS OF EEG

The human Electroencephalogram (EEG) is usually recorded from electrodes attached to the scalp using high amplifiers, which are usually coupled to the scalp electrodes. The amplified signals are written out on paper via a polygraph, which contains typically 8 to 16 channels. Normal subjects usually exhibit alpha, beta, theta and delta activities, while abnormal activity may be manifested by a slowing and decrease in amplitude of EEG, increase in the EEG frequency, and the presence of sudden EEG discharges (paroxysmal activity) different from the background either in frequency content or amplitude or pattern.

The EEG is a powerful tool for the diagnosis of neurological disorders. Since its discovery, the EEG has been used for the diagnosis of epilepsy, for trauma assessment, for sleep research, and for the analysis of higher brain functions. The EEG is highly dependent upon the availability of high quality instrumentation, and almost from the beginning, automated methods of signal qualification have been applied. One of the primary goals is to help the Encephalographer (EEGer) in the time consuming task of quantifying signal that appears to the eye as a low information content background intermixed with either bursts of rhythmic activity with different frequencies (the EEG rhythms) or short transients of clinical significance (such as spikes). In spite of years of research to produce universal automated detection methods, success has been achieved only in specific areas. Accomplishments include automatically sleep staging with a high degree of accuracy; counting spikes and wave complexes, and monitoring in intensive care units. However clinicians still rely on visual analysis for clinical applications.

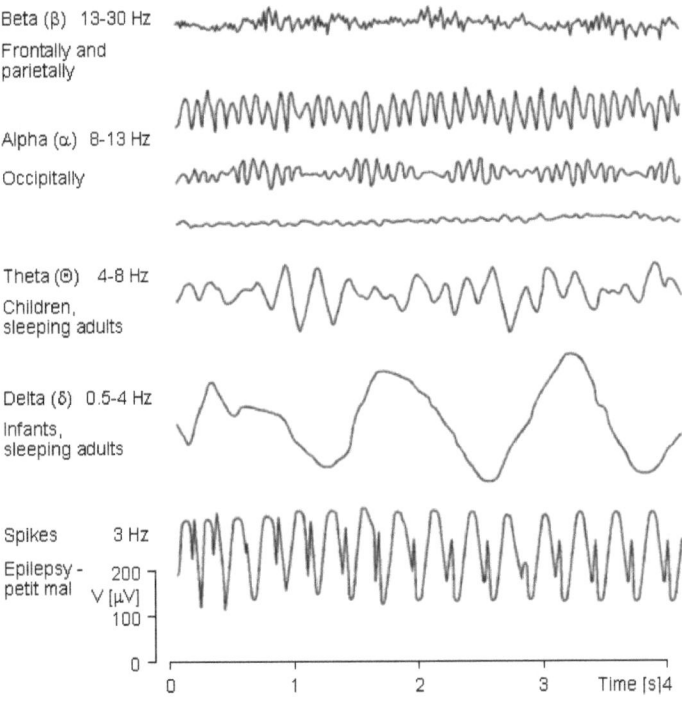

Figure 1.1 EEG Waveform

The human eye-brain can be trained to recognize ostensibly defined patterns in multi-channel EEG recordings. However, ostensive definitions are not readily disseminated. A description of a mental image by words is normally poor and lengthy. What is needed in a clinical practice is a way of exploring the great pattern recognition of a human visual system and enhancing the efficiency of the visual data communication.

Computers can bring quantification to EEG analysis in the form of precise Measurements (micro volt and millisecond precision), but at this time they cannot always use the measured data to identify clinically significant features. All these aspects lead us to approach the use of computers in EEG research from a slightly different angle. We are also researching the design of computer-based environments that will help the doctor in the visual clinical assessment of multi-channel EEG recordings, and the engineer in the design of better detectors.

It is widely accepted that the information available to the physician about his patient and about medical relationships in general is inherently uncertain. Nevertheless, the physician is still quite capable of drawing conclusions, though approximate, from this information. The novel attempt in this project is to provide a formal model of this process using a mathematical approach in implementing the model in the form of a computerized diagnostic system. The two contrasting and complementary approach include onset indication by aggregation analysis and optimized classification of the risk level of epilepsy patients.

1.2   EPILEPSY TYPES AND SYMPTOMS

While many types of repetitive behavior may represent a neurological problem, a doctor needs to establish whether or not they are seizures[4].

1.2.1   GENERALIZED SEIZURES

- All areas of the brain (the cortex) are involved in a generalized seizure. Sometimes these are referred to as grand mal seizures.
- The person experiencing such a seizure may cry out or make some sound, stiffen for several seconds to a minute and then have rhythmic movements of the arms and legs. Often the rhythmic movements slow before stopping.
- Eyes are generally open.

- The person may appear to not be breathing and actually turn blue. This may be followed by a period of deep, noisy breathes.
- The return to consciousness is gradual and the person may be confused for quite some time -- minutes to hours.
- Loss of urine is common.
- The person will frequently be confused after a generalized seizure.

1.2.2 PARTIAL OR FOCAL SEIZURES

- Only part of the brain is involved, so only part of the body is affected. Depending on the part of the brain having abnormal electrical activity, symptoms may vary[5].
- If the part of the brain controlling movement of the hand is involved, then only the hand may show rhythmic or jerky movements.
- If other areas of the brain are involved, symptoms might include strange sensations like a full feeling in the stomach or small repetitive movements such as picking at one's clothes or smacking of the lips.
- Sometimes the person with a partial seizure appears dazed or confused. This may represent a complex partial seizure. The term *complex* is used by doctors to describe a person who is between being fully alert and unconscious.

1.2.3 ABSENCE OR PETIT MAL SEIZURES

- These are most common in childhood.
- Impairment of consciousness is present with the person often staring blankly.
- Repetitive blinking or other small movements may be present.
- Typically, these seizures are brief, lasting only seconds. Some people may have many of these in a day.

1.3 MATHEMATICAL APPROACH IN MEDICAL DIAGNOSIS

Mathematics is one of the most useful and fascinating divisions of human knowledge. The most important skills in mathematics are careful analysis and clear reasoning, entirely based on logic. Starting with widely accepted statements, mathematics can be used to draw logical conclusions and develop complete systems based on such conclusions. The two forms of mathematics, namely *pure* and *applied mathematics,* do not have a clear preset boundary between them. Where pure mathematics seeks to advance mathematical knowledge, applied mathematics seeks to develop mathematical techniques for use in science and other fields.

Mathematics is an essential part of any scientific study. It provides a plethora of techniques to analyze, quantify and qualify the scientific data. The same can be said about application of mathematical techniques in medical diagnosis. Frequency – time analysis of signals is possible with the aid of Fourier Transforms and likewise, other analytical methods can be used in obtaining different characteristics of the various signals originating in different parts of the body, helpful in determining defects and other inherent pathologies.

Precision exists only through abstraction. Abstraction may be defined as the ability of human beings to recognize and select the relevant properties of real world phenomena and objects. This leads to the construction of conceptual models defining abstract classes of phenomena and objects. However, in actuality every real-world phenomenon and object is of course unique. Abstract models of real-world phenomena and objects such as mathematical structures (circle, point, etc.), equalities ($a=b + c$), and propositions (yes, no) are artificial constructs. They represent ideal structures, ideal equalities, and ideal positions. Nevertheless, despite these caveats, abstraction forms the basis of human thought, and human knowledge is its result.

1.4 EEG SIGNALS FOR EPILEPSY DETECTION

Epileptic seizures result from a temporary electrical disturbance of the brain. Sometimes seizures may go unnoticed, depending on their presentation, and sometimes may be confused with other events, such as a stroke, which can also cause falls or migraines. Approximately one in every 100 persons will experience a seizure at some time in their life. Unfortunately, the occurrence of an epileptic seizure seems unpredictable and its process is very little understood. Since its discovery by R.Caton, the Electroencephalogram (EEG) has been the most utilized signal to clinically assess brain activities[7][8][9]. Twenty –five percent of the world's 50 million people with epilepsy have seizures that cannot be controlled by any available treatment. The need for new therapies, and success of similar devices to treat cardiac arrhythmias, has spawned an explosion of research into algorithms for use in implantable therapeutic devices for epilepsy. Most of these algorithms focus on either detecting unequivocal EEG onset of seizures or on quantitative methods for predicting seizures in the state space, time, or frequency domains that may be difficult to relate to the Neuro physiology of epilepsy. Between seizures, the EEG of a patient with epilepsy may be characterized by occasional epileptic form transients-spikes and sharp waves. EEG patterns have shown to be modified by a wide range of variables including biochemical, metabolic, circulatory, hormonal, neuro electric and behavioral factors [10].

Exploring various analytical approaches, both linear and non linear methods to process data from medical database is meaningful before deciding on the tool that will be most useful, accurate, and relevant for practitioners. For example, assigning a new patient to a particular outcome class is a classification problem commonly described as "pattern recognition", "discriminant analysis", and "supervised learning". In the past, the Encephalographer, by visual inspection was able to qualitatively distinguish normal EEG activity from localized or generalized abnormalities contained within relatively long EEG records. The different types of epileptic seizures are characterized by different EEG waveform patterns. With real-time monitoring to detect epileptic seizures gaining widespread recognition, the advent of computers has made it possible to effectively apply a host of methods to quantify the changes occurring based on the EEG signals. One of them is a classification of risk level of epilepsy by using Fuzzy techniques. The recognition of specific waveforms and features in the Electroencephalogram (EEG) for classification of epilepsy risk levels has been the subject of much study [20].

Electroencephalography is a well-established clinical procedure, which can provide information pertinent to the diagnosis of a number of brain disorders (e.g., epilepsy or brain tumors). However, despite its widespread use, it is one of the last routine clinical procedures to be fully automated. Analysis of the electroencephalogram (EEG) includes the detection of patterns and features characteristic of abnormal conditions. For example, Asymmetries in the amplitude or frequency of background activity suggest a lesion, while the presence of epileptiform activity supports a clinical diagnosis of epilepsy. Over half the EEG referrals relate to epilepsy, with the EEG being the most useful procedure in its diagnosis.

Recording the EEG during a seizure is particularly helpful in determining whether a patient has epilepsy. Because seizures usually occur infrequently and unpredictably, obtaining such recording might require an EEG extending over several days (long-term EEG monitoring). Techniques have been developed for the automated detection of petitmal seizures and grand mal seizures, which have proven relatively successful.

Between seizures, the EEG of a patient with epilepsy may be characterized by occasional epileptiform transients (spikes and sharp waves) and, consequently, relatively short recording can still be useful in the diagnosis of epilepsy. A routine recording typically lasts 20-30 minutes, during which some 4 minutes of paper record are produced[. An Electroencephalographer (EEGer) detects epileptiform transients by visual inspection of the recording, which requires considerable skill and is time consuming. Hence, automation of this process could save time increase objectivity and uniformity, and enable quantification for research studies.

Automated detection of epileptiform transients has two primary areas of clinical application. The first is in long term EEG monitoring, where it acts essentially as a daily reduction process. A segment of EEG is recorded only when a transient is detected and all segments are reviewed by an EEGer. Thus, the goal is to detect a high proportion of epileptiform activity while minimizing false detection. The second area is in routine clinical recordings where, major objective is to minimize the visual inspection process as far as epileptiform transients are concerned. In this case it is important not to precipitate a misdiagnosis of epilepsy and, therefore, the aim is to eliminate false detections while detecting a satisfactory proportion of epileptiform transients [21].

Spikes and sharp waves are defined as transients clearly distinguished from background activity with pointed peaks at conventional paper speeds. Spikes are defined having durations of 20-70 ms, while sharp waves have durations of 70-200 ms. No distinction is made between spikes and sharp waves and, therefore, they are collectively termed epileptiform transients. Due to the variety of morphologies of epileptiform transients and similarities to waves which are part of the background activities and due to artifacts (i.e., extra cerebral potentials from muscles, eyes, heart, electrodes, etc.), the detection of epileptiform activity in the EEG is far from straightforward.

Several techniques have been applied to the detection of epileptiform activity in the EEG.

These include:

a) Template matching, where a detection is made whenever the value relation of the EEG with a template exceeds a threshold

b) Parametric methods, where a detection is made when the difference between the EEG and its predicted value used on the assumption that the background is stationary exceeds a threshold

c) Mimetic methods, where one or more parameters of each wave are calculated and threshold.

d) Syntactic methods, where detections are based on the presence of a structural combination of structures

e) Artificial neural networks trained to detect epileptic waveform transients and

f) Expert systems, which detect epileptiform activity by mimicking the knowledge and reasoning of the EEGer.

Most of these systems are in the developmental stage, and those in clinical use are restricted to long-term EEG monitoring with all detections being reviewed by an EEGer.

Due to a high number of false detections these systems cannot perform satisfactorily in the routine EEG setting.

It is generally accepted that the only way to separate epileptiform from non-epileptiform waves is to make use of a spatial and temporal context. Several groups are implementing this approach in an effort to minimize false detections. Glover et al. have developed a system that relies on a wide spatial context, with 12 EEG channels being analyzed together with additional contextual information provided by EKG, EOG, and EMG signals. Conversely, the system developed by Gotman and Wang implements as extra temporal context, where sections of EEG are classified into one of the five states (active wakefulness, quiet wakefulness, synchronized EEG, phasic EEG, or slow-wave EEG) before independent rules are applied to reject non-epileptiform activity.

A new system has been developed, that makes considerable use of spatial and temporal contextual information. This system is proven to be particularly successful at rejecting non-epileptiform activity during awake and resting EEG's. It uses a mimetic the method to detect candidate transients, which are subsequently trimmed or rejected as epileptiform by an expert system. The system integrates both spatial and temporal contextual information to detect definite and probable epileptiform activities and reject non-epileptiform waves. Preliminary results state that this system should be capable of performing routine clinical EEG setting.

1.5 ORGANIZATION OF THE PROJECT REPORT

The block diagram of KNN Classifier and K- Means Clustering based epilepsy risk level classifier is shown in figure1.1. This is accomplished as:

1. Dimensionality reduction of the EEG dataset using Power Spectral estimation of the EEG signals.
2. Epilepsy Risk level Classification by applying KNN Classifier.
3. Epilepsy Risk level Classification by applying K- Means Clustering and comparison between above technique.

The Electroencephalogram signals from epileptic patients are to be collected from hospitals. The collected waveforms are digitized to get the discrete values of the features of the EEG waveform. The number of samples is very large. Hence it is mandatory to do dimensionality reduction in order to lower the computational complexity. The power spectral density of the EEG signals is estimated. Only the maximum Power spectral values are chosen as parameters.

To these values KNN Classifier and K-Means Clustering algorithm are applied to classify the epilepsy risk level. The various performance parameters like Performance index, Sensitivity, Specificity, Quality value, False alarm and Missed classification.

CHAPTER 2
MATERIALS AND METHODS

The EEG is recorded by placing electrodes on the scalp according to the International 10-20 system. Sixteen channels of EEG are recorded simultaneously for both referential montages, where all electrodes are referenced to a common potential like ear, and bipolar montages, where each electrode is referenced to an adjacent electrode. The EEG recording points on the scalp are illustrated in figure 1.2. Recordings are made while the patient is awake but resting and include periods of eyes open, eyes closed, hyperventilation and photoic stimulation. Amplification is provided by an EEG machine (Siemens Minograph Universal).

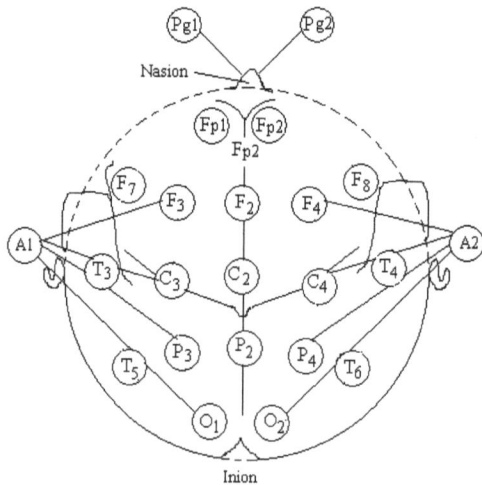

Figure 2.1 EEG Recording by 10-20 system

Before placing the electrodes, the scalp is cleaned, lightly abraded and electrode paste is applied between the electrode and the skin. By means of this application of electrode paste, the contact impedance is less than 10 KΩ. Generally disk like surface electrodes are used. In some cases, needle electrodes are used to pick up the EEG signals. The signals are recorded with the speed of 30 mm/s. The obtained signals are filtered by notch filter (low pass filter - 5Hz, high pass filter - 75Hz).

2.1 ACQUISITION OF EEG DATA

The EEG data used in the study were acquired from ten epileptic patients who had been under the evaluation and treatment in the Neurology department of Sri Ramakrishna Hospital, Coimbatore, India. A paper record of 16 channel EEG data is acquired from a clinical EEG monitoring system through 10-20 international electrode placing method. The EEG signal was band pass filtered between 0.5 Hz and 50Hz using five pole analog Butter worth filters to remove the artifacts. With an EEG signal free of artifacts, a reasonably accurate detection of epilepsy is possible; however, difficulties arise with artifacts. This problem increases the number of false detection that commonly plagues all classification systems. With the help of Neurologist (Golden standard with 100% sensitivity &100% specificity), we had selected artifact free EEG records with distinct features. These records were scanned by Umax 6696 scanner with a resolution of 600dpi

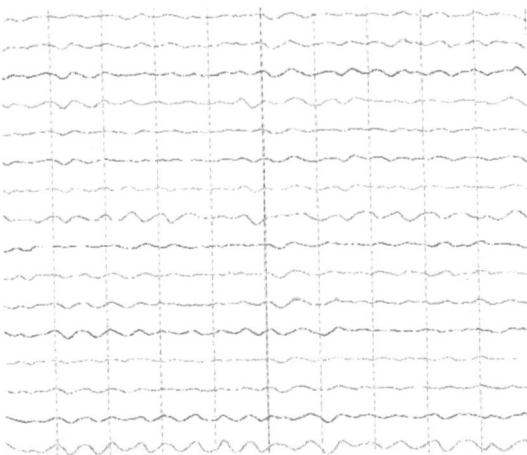

Figure 2.2 Epileptic EEG Signal Waveform of Patient

The EEG is broken down into sections or epochs, for the purpose of feature extraction. An epoch of 2.0 s is used for the following reasons:

1) It is long enough to capture the main statistical characteristics of the EEG and short enough to capture the evolution of seizures

2) The EEG being digitized at a sampling rate of 200 Hz an epoch of 2s contains 400 samples, which is a convenient length for computation. The software for analyzing the EEG data was implemented using C++ programming and Mat lab 7.2. Waveforms of normal and abnormal data are plotted and studied.

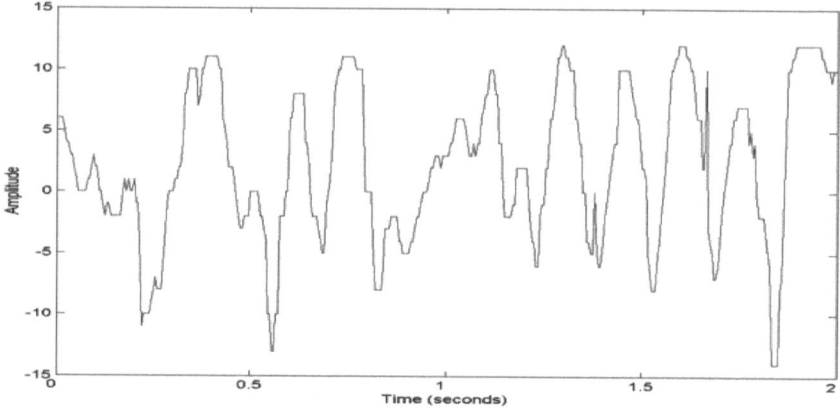

Figure 2.3 Sample 2-second epoch

A group of twenty patients with known clinical findings of epileptic seizure is undertaken for classifications of level of epilepsy risk.

2.2 DIMENSIONALITY REDUCTION

The pixels of the bmp files are converted to x and y coordinates where the y coordinate represents the signal amplitude value. The signals are reconstructed with the following scaling factor:

X-axis: 60mm = 2seconds

Y-axis: 1mm = 70μV

The X-axis of the scaled image is set to a width of 400 pixels so that each pixel represents a sampled amplitude value. These amplitude values are found using graphics programming in C++ and are written to a file.

The final risk level classification is done by KNN Classifier and K- Means clustering. The twenty patients are labeled into nine groups based upon their EEG signals and clinical observation. The patients are assigned and labeled in groups based on the severity index or target values of the frequency occurrence of epileptic seizures. The patients who in the G_1 group are considered to be more vulnerable one and G_9 group s lest disturbed.

Table 2.1 Target Values for Groups

GROUP	PATIENT NUMBER	TARGET VALUE
G_1	1,2,11,20	0.85 (High Risk)
G_2	3,7,14,18	0.65
G_3	4,13	0.45
G_4	6,12	0.35
G_5	5,8,19	0.25
G_6	15,17	0.20
G_7	16	0.15
G_8	10	0.1
G_9	9	0.05(Low Risk)

Figure 2.4 depicts the flow diagram of the epilepsy risk level classification system as shown below.

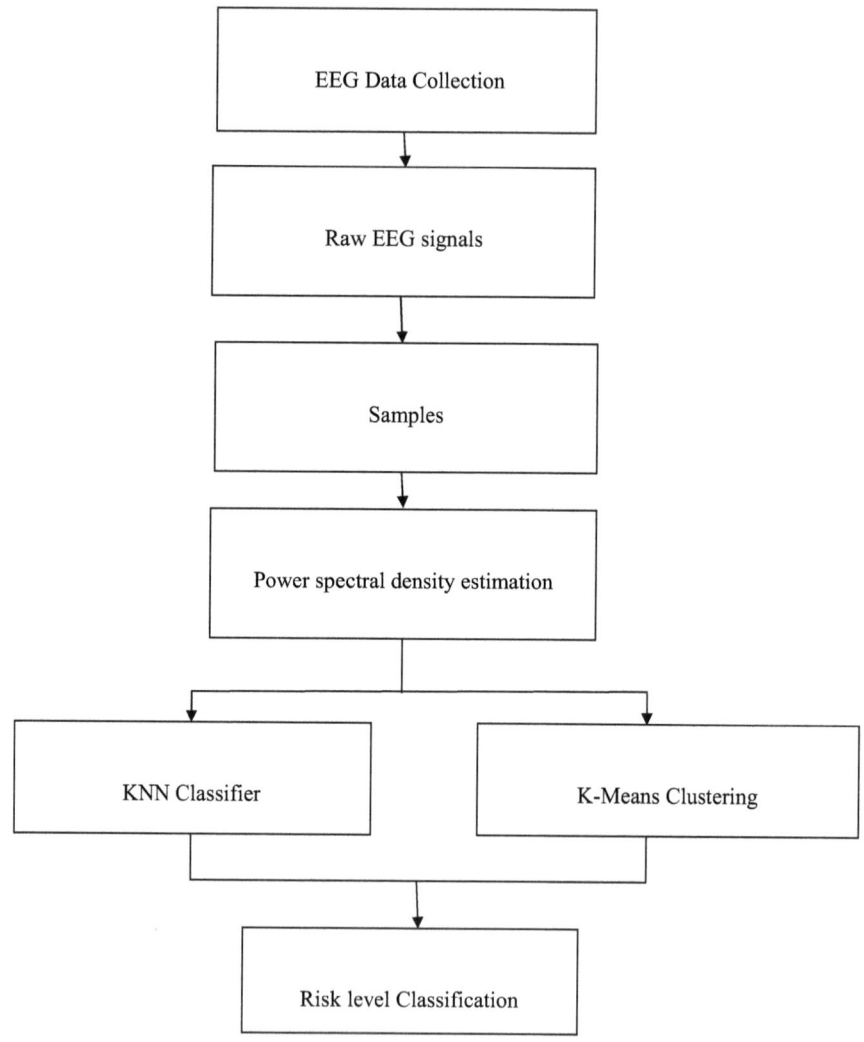

Figure 2.4 Flow diagram of the Epilepsy Risk Level Classification System

2.2.1 POWER SPECTRAL DENSITY

Power Spectral Density (PSD) is the frequency response of a random or periodic signal. It tells us where the average power is distributed as a function of frequency. The PSD is deterministic, and for certain types of random signals is independent of time[1]. This is useful because the Fourier transform of a random time signal is itself random, and therefore of little use calculating transfer relationships (i.e., finding the output of a filter when the input is random). The PSD of a random time signal x(t) can be expressed in one of two ways that are equivalent to each other.

1. The PSD is the average of the Fourier transform magnitude squared, over a large time interval

$$S_x(f) = \lim_{T \to \infty} E\left\{ \frac{1}{2T} \left| \int_{-T}^{T} x(t) e^{-j2\pi ft} dt \right|^2 \right\} \qquad 2.1$$

2. The PSD is the Fourier transform of the auto-correlation function.

$$S_x(f) = \int_{-T}^{T} R_x(\tau) e^{-j2\pi ft} dt \qquad 2.2$$

$$R_x(\tau) = E\{x(t) x^*(t+\tau)\} \qquad 2.3$$

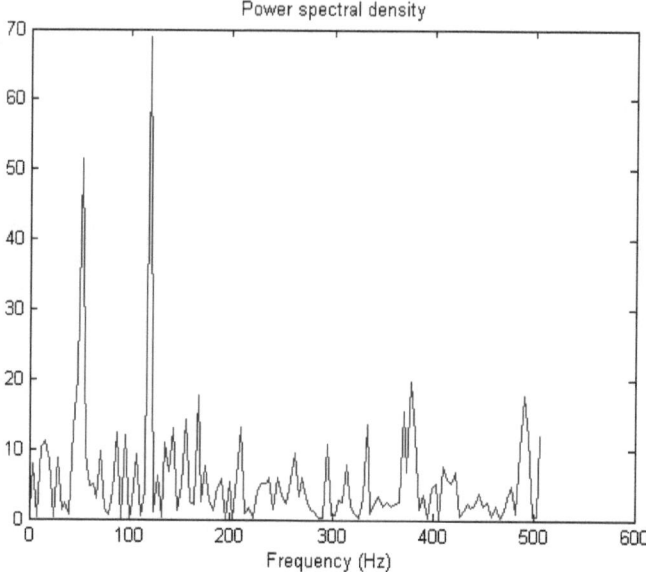

Figure 2.5 Power spectral density of a signal

Power spectra can be computed for the entire signal at once (a "periodogram") or periodograms of segments of the time signal can be averaged together to form the "power spectral density"[26].

2.2.2 PERIODOGRAM

The periodogram computes the power spectra for the entire input signal:

$$\text{Periodogram} = \frac{|abs(F(\text{signal}))|^2}{N}$$

2.4

where F(signal) is the Fourier transform of the signal, and N is the normalization factor, which Igor's DSPPeriodogram operation defaults to the number of samples in the signal.

The calculation of the periodogram is improved by spectral windowing, and Igor's DSPPeriodogram operation supports the same windows as the FFT operation does.

The result of the periodogram is often normalized by a multiplication factor to make the result satisfy Parseval's Theorem:

$$\frac{1}{N}\sum_{i=0}^{N-1}|\text{FFT amplitude}[i]|^2 = \sum_{i=0}^{N-1}|\text{data magnitude}[i]|^2$$

2.5

which presumes the two-sided frequency-domain FFT result is computed from the time-domain signal data, and where N is again the number of time-domain values in the signal.

Normalization of the periodogram result to meet this criterion follows several different conventions in the literature, (and depends on the average power of any spectral windowing function and also on whether the periodogram is one- or two-sided), so the DSPPeriodogram operation allows the user to specify the desired normalization using the /NOR parameter. When using a window function, the amount of power in the signal is reduced.

2.2.3 WINDOW FUNCTION

Window function is a mathematical function that is zero valued outside of some chosen period. When another function or waveform or data sequence is multiplied by a window function, the product is also zero - valued outside the period; all that is left is the part where they overlap the observation through the window. It is simple to apply and understand. In this method the frequency of a filter, HD(w) and the corresponding impulse response, hD(n), are related by the inverse Fourier transform:

$$h_D(n) = \frac{1}{2\pi}\int_{-\pi}^{\pi} H_D(w) e^{jwn} dw$$

2.6

where HD(w) is frequency response of a filter and hD(n) is corresponding impulse response. Here, HD(w) can be obtained from hD(n) by evaluating the inverse Fourier transform. The truncation of hD(n) to a length M-1 is equivalent to multiplying hD(n) by a rectangular window.

The types of windows are

- Rectangular window
- Hamming window
- Hanning window
- Bartlett window
- Blackman turkey window

The features of window functions are given below.

Table 2.2 Features of various windowing techniques

Window function	Transition Width (Hz)	Passband Ripple (dB)	Main lobe relative to side lobe (dB)	Stopband Attenuation (dB)	Time domain sequence
Rectangular window	0.9/N	0.7416	13	21	1
Hamming window	3.3/N	0.0546	41	53	$0.54 + 0.46 \cos(2\Pi n/N)$
Hanning window	3.1/N	0.0646	31	44	$0.5[1-\cos(2\Pi n/N -1)]$
Bartlett window	0.89k2Π/N				
Blackman - Turkey window	5.5/N	0.0017	57	75	$0.42 - 0.5 \cos(2\Pi n/N - 1) + 0.08 \cos(4\Pi n/N -1)$

2.2.4 RECTANGULAR WINDOW

Rectangular window is defined as:

$$w(n) = \begin{cases} 1, & n = 0,1,2,\dots.M-1 \\ 0 & \text{otherwise} \end{cases} \qquad 2.7$$

And unit impulse response of rectangular window is given below

$$h(n) = h_D(n)w(n)$$
$$h(n) = \begin{cases} h_D(n), & n = 0,1,2,\dots.M-1 \\ 0 & \text{otherwise} \end{cases} \qquad 2.8$$

Frequency domain function in representation of window function is

$$W(w) = \sum_{n=0}^{M-1} w(n) e^{-jwn} \qquad 2.9$$

The Rectangular window has admirable resolution characteristics for sinusoids of comparable strength. When the signal is harmonically related to the window length in first harmonic frequency corresponding to a record length, the signal appears periodic and infinite and is faithfully reproduced even by the rectangular window. The individuality of it play a significant role in establishment of the resulting frequency response of the finite impulse response filter obtained by truncation hD(n) to length M. The undesirable effects are best alleviated by the use of window that do not contain abrupt discontinuities in their time Domain characteristics and have likewise low side lobes in their frequency domain characteristics.

For the same value of M for both Rectangular and Hamming window and other windows, the width of the main lobe is also wider for, these windows compared to the rectangular window.

The Fourier transform of rectangular window

$$W(w) = \sum_{n=0}^{M-1} e^{-jwn}$$

$$= \frac{1-e^{-jwM}}{1-e^{-jw}} = e^{-jw(M-1)/2} \frac{\sin(wM/2)}{\sin(w/2)} \qquad 2.10$$

Each sample corresponds to the instantaneous amplitude values of the signal, totalling 400 values for an epoch. Each channel has 400 samples of EEG signals per epoch and four such epochs of data forms a bin. There are sixteen such a bins are available per patient. The data volume for a patient is around 25,600 samples. Hence this large amount of data necessitates the dimensionality reduction technique level for processing the EEG signal.

Dimensionality Reduction (DR) is a pre-processing step that reduces the dimension of the EEG data. Conventional decomposition methods such as Singular Value Decomposition (SVD), Principle Component Analysis (PCA), and Independent Component Analysis (ICA) reduce the dimensions of the data by isolating a set of features or electrodes that comply with some certain criteria regardless of their impact on classification. In addition, these techniques require the entire trials of a recording session of subject for the reduction process[2][13]. To overcome the drawbacks of traditional methods we opt for other dimensionality reduction technique.

In this work we have used power spectral density as dimensionality reduction [15]. The power spectral densities for the whole twenty patients' samples are estimated. We have implemented periodogram method for power spectral density estimation. Rectangular window of size 20 is applied. The power spectral density (PSD) has been extracted as features. Only maximum PSD values are chosen as parameters. To these values KNN Classifier and K-Means clustering is applied.

CHAPTER 3
KNN CLASSIFIER

3.1 OVERVIEW OF *K*-NEAREST NEIGHBOR ALGORITHM

The KNN is one of prospective statistical classification algorithms used for classifying objects based on closest training examples in the feature space. It is a lazy learning algorithm where the KNN function is approximated locally and all computations are deferred until classification [16]. No actual model or learning is performed during the training phase, although a training dataset is required, it is used solely to populate a sample of the search space with instances whose class is known, for this reason, this algorithm is also known as lazy learning algorithm. It means that the training data points are not used to do any generalization and all the training data is needed during the testing phase. When an instance whose class is unknown is presented for evaluation, the algorithm computes its K closest neighbors, and the class is assigned by voting among those neighbors. In KNN algorithm, training phase is very fast but testing phase is costly in terms of both time and memory.

The k-nearest-neighbor classifier (KNN Classifier for short) is one of the most basic classifiers for pattern recognition or data classification [34]. The principle of this method is based on the intuitive concept that data instances of the same class should be closer in the feature space. As a result, for a given data point x of unknown class, we can simply compute the distance between x and all the data points in the training data, and assign the class determined by the K nearest points of x. Due to its simplicity, KNNC is often used as a baseline method in comparison with other sophisticated approaches in pattern recognition [35].

Given a training dataset of n points with their desired class, as shown below:

$$\{(x_i, y_i), (x_2, y_2), \ldots, (x_n, y_n)\}$$

where (x_i, y_i) represent data pair i, with x_i as the feature vector and y_i as the corresponding target class. Then for a new data point x, the most likely class should be determined by KNNC (k = 1 in this case), as follows:

$$nnc(x,1) = y_{p}, p = \operatorname{argmin}_{i} \|x - x_i\|2.$$

The preceding equation uses the nearest neighbor to determine the class. Alternatively, we can have K nearest neighbors to determine the class by voting

The KNN algorithm consists of two phase

- Training phase
- Classification phase.

In training phase, the training examples are vectors (each with a class label) in a multidimensional feature space. In this phase, the feature vectors and class labels of training samples are stored.

In the classification phase, K is a user-defined constant, a query or test point (unlabelled vector) is classified by assigning a label, which is the most recurrent among the K training samples nearest to that query point. In other words, the KNN method compares the query point or an input feature vector with a library of reference vectors, and the query point is labeled with the nearest class of library feature vector. This way of categorizing query points based on their distance to points in a training data set is a simple, yet an effective way of classifying new points.

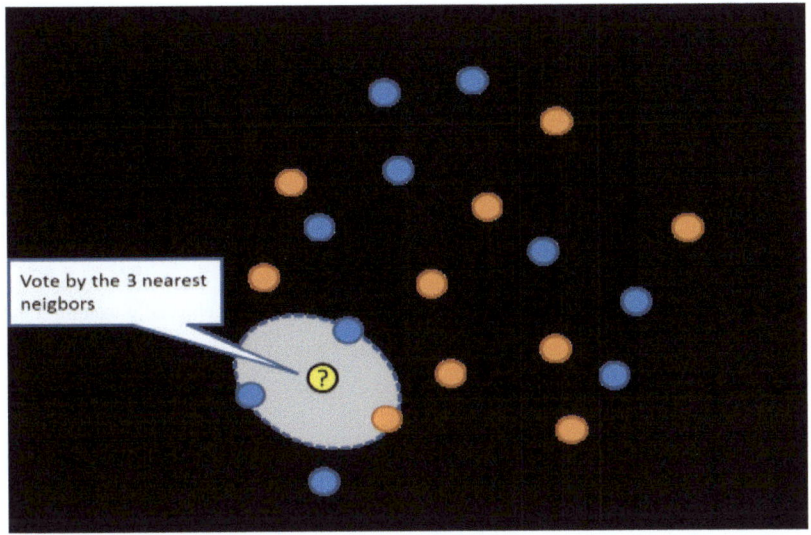

Figure 3.1 KNN Classifier

3.2   PARAMETER *K*

One of the advantages of the KNN method in classifying the objects is that it requires only few parameters to tune: K and the distance metric, for achieving sufficiently high classification accuracy. Thus, in KNN based implementations the best choice of K and distance metric for computing the nearest distance is a critical task. Generally, larger values of K reduce the effect of noise on the classification, but make boundaries between classes less distinct. The special case where the class is predicted to be the class of the closest training sample (i.e. when $K = 1$) is called the nearest neighbor algorithm. In binary classification problems, it is helpful to choose K to be an odd number as it avoids tied votes. Thus, the value of K is defined in such a way that it produces the highest correct classification rate. In this work the different values of K which have been tested are 1, 3, 5, 7 and 9. Further, the different distance metrics which are used in this work are Euclidean distance, City Block and Correlation. The brief explanation of these distance metrics is as follows:

3.3 DISTANCE METRICS

Given an $[mx \times n]$ data matrix X, which is treated as $[mx(1 \times n)]$ row vectors $x_1, x_2, \ldots, x_{mx}$, and $[my \times n]$ data matrix Y, which is treated as $[my(1 \times n)]$ row vectors $y_1, y_2, \ldots, y_{my}$, the various distances between the vector x_s and y_t are defined as follows:

3.3.1 EUCLIDEAN DISTANCE METRIC (EU)

This is the most usual way of computing a distance between two objects. It examines the root of square differences between coordinates of a pair of objects and is defined using the following equation.

$$d_{st} = \sqrt{\sum_{j=1}^{n}(x_{sj} - y_{tj})^2} \qquad 3.1$$

3.3.2 CITY BLOCK DISTANCE METRIC (CB)

It is based on Taxicab geometry, first considered by Hermann Minkowski in the 19th century, is a form of geometry in which the usual metric of Euclidean geometry is replaced by a new metric in which the distance between two points is the sum of the absolute differences of their coordinates defined using the following equation:

$$d_{st} = \sum_{j=1}^{n}|x_{sj} - y_{tj}| \qquad 3.2$$

The City Block distance metric is also known as Manhattan distance, boxcar distance, absolute value distance. It represents distance between points in a city road grid. While the Euclidean corresponds to the length of the shortest path between two points (i.e. "as the crow flies"), the City Block distance refers to the sum of distances along each dimension (i.e. "walking round the block").

3.3.3 CORRELATION DISTANCE METRIC (CO)

It is one minus the sample correlation between points (treated as sequences of values) and is defined using the following equation:

$$d_{st} = 1 - \frac{(x_s - \bar{x}_s)(y_t - \bar{y}_t)'}{\sqrt{(x_s - \bar{x}_s)(x_s - \bar{x}_s)'}\sqrt{(y_t - \bar{y}_t)(y_t - \bar{y}_t)'}} \qquad 3.3$$

where

$$\bar{x}_s = \frac{1}{n}\sum_j x_{sj} \quad \text{and} \quad \bar{y}_t = \frac{1}{n}\sum_j y_{sj}$$

It is important to mention here that the performance of classifiers is usually dependent upon the value of K and distance metric. In this work these values are evolved through cross-validation, which involves the determination of classification accuracy for multiple partitions of the input samples used in training. The cross-validation is mainly used in settings where the goal is prediction, and one wants to estimate how accurately a predictive model will perform in practice.

In general, one round of cross-validation involves partitioning a sample of data into complementary subsets, performing the analysis on one subset called the training set, and validating the analysis on the other subset called the testing set or validation set. Further, to reduce variability, multiple rounds of cross-validation are performed using different partitions, and the validation results are averaged over the rounds.

In k-fold cross-validation, the original sample is randomly partitioned into k subsamples. Of the k subsamples, a single subsample is retained as the validation data for testing the model, and the remaining ($k - 1$) subsamples are used as training data. The cross-validation process is then repeated k-times (the folds), with each of the k subsamples used exactly once as the validation data. The k results from the folds then can be averaged to produce a single estimation.

3.4 KNN CLASSIFIER ALGORITHM

- Compute the test point's distance from each training point
- Sort the distances in ascending (or descending) order
- Use the sorted distances to select the K nearest neighbors
- Use majority rule (for classification) or averaging (for regression).

The dimensionality reduced power spectral values are given as input to the KNN Classifier. Here the Euclidean distance metric is used for similarity measurement between the various points. Other distance metrics are city block, hamming distance. We opt for the standard Euclidean distance. The Euclidean distance between two points is given by

$d(x, y) = sqrt(x^2+y^2)$

Target values for each patient are fixed as per the table 2.2. The efficiency of the KNN Classifier is analyzed through various performance parameters. The performance parameters are namely Performance index, false alarm, Missed classification, Sensitivity, Specificity and Quality value.

The values obtained are tabulated below. It has been observed the quality value obtained is low. We have implemented the KNN Classifier to these PSD values and its performance is compared with K-Means Clustering. The next chapter details about it.

Table 3.1 Performance parameters of KNN Classifier

Patient	PC (%)	MC (%)	FA (%)	PI (%)	Se (%)	Sp (%)	Av Detection (%)	Time delay (Second)	Qv
1	88.89	5.55	5.55	87.14	94.44	94.44	94.34	2.10	19.17
2	88.89	0	08	87.29	88.89	100	94.44	1.77	18.30
3	93.06	2.77	4.16	92.49	95.83	97.23	96.53	2.027	20.61
4	84.72	12.49	2.77	78.85	90.27	87.50	92.36	2.44	19.13
5	83.33	6.94	9.72	79.42	97.23	93.05	91.66	2.08	17.45
6	77.78	19.44	2.77	69.42	87.50	80.56	87.55	2.72	16.94
7	80.55	6.94	12.49	74.47	80.55	93.05	90.27	2.16	17.08
8	80.55	0	19.42	73.66	80.56	100	90.27	1.61	16.57
9	76.39	0	23.58	63.63	76.39	100	88.18	1.69	16.76
10	83.33	15.27	1.38	78.66	98.61	84.72	91.67	2.58	18.62

11	86.11	0	13.86	83.83	86.11	100	93.05	1.72	17.98
12	76.38	0	23.58	72.52	76.38	100	88.19	1.52	15.98
13	86.11	8.33	5.53	83.68	94.44	91.67	93.05	2.22	18.87
14	69.45	9.72	20.83	55.88	79.17	90.28	84.72	1.97	14.73
15	72.23	22.22	5.53	60.02	94.46	97.78	86.11	2.77	15.51
16	83.33	15.27	1.38	77.15.	98.61	84.72	91.66	2.57	18.97
17	95.83	0	4.16	5.52	95.83	100	97.91	1.1	21.94
18	87.49	6.92	5.53	85.26	99.44	93.05	93.75	2.16	19.42
19	83.33	11.06	5.53	80.07	94.44	88.86	91.66	2.33	17.
20	88.89	0	11.08	87.29	88.89	100	94.44	1.77	18.30
Average	83.33	7.14	9.50	78.31	90.48	92.84	87.09	2.10	18.02

CHAPTER 4
K- MEANS CLUSTERING

4.1 INTRODUCTION TO K-MEANS CLUSTERING

k-means is one of the simplest unsupervised learning algorithms that solve the well known clustering problem[18]. The procedure follows a simple and easy way to classify a given data set through a certain number of clusters (assume k clusters) fixed prior. The main idea is to define k centers, one for each cluster[30]. These centers should be placed in a cunning way because of different location causes different result. So, the better choice is to place them as much as possible far away from each other. The next step is to take each point belonging to a given data set and associate it to the nearest center. When no point is pending, the first step is completed and an early group age is done. At this point we need to re-calculate k new centroids as barycenter of the clusters resulting from the previous step. After we have these k new centroids, a new binding has to be done between the same data set points and the nearest new center[19]. A loop has been generated. As a result of this loop we may notice that the k centers change their location step by step until no more changes are done or in other words centers do not move any more. Finally, this algorithm aims at minimizing an objective function know as squared error function given by:

$$SSE = \sum_{i=1}^{C} \sum_{j=1}^{C_i} (||X_i - Y_j||)^2 \qquad 4.1$$

where,

'$||X_i - Y_j||$' is the Euclidean distance between X_i and Y_j ; 'C_i' is the total number of pattern points in i^{th} cluster; 'C' is the number of cluster centroids.

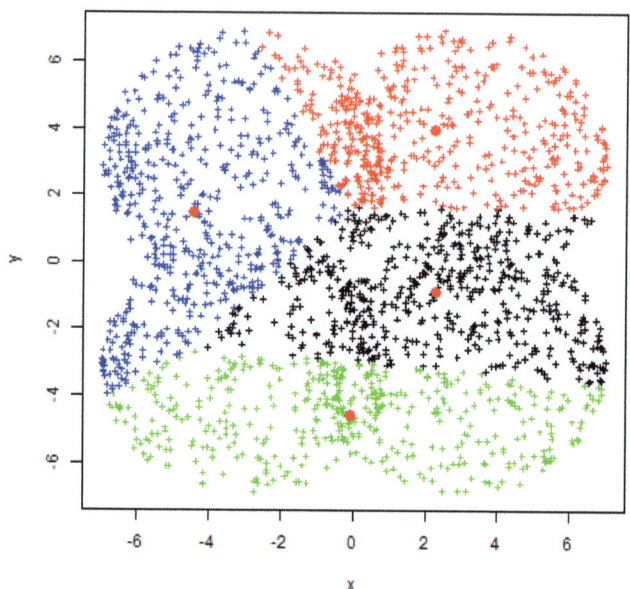

Figure 4.1 K-Means clustering working

4.2 ALGORITHMIC STEPS FOR K-MEANS CLUSTERING

Let $X = \{x_1, x_2, x_3, \ldots, x_n\}$ be the set of data points and $V = \{v_1, v_2, \ldots, v_c\}$ be the set of centers.

1) Randomly select *'c'* cluster centers.

2) Calculate the distance between each data point and cluster centers.

3) Assign the data point to the cluster center whose distance from the cluster center is minimum of all the cluster centers..

4) Recalculate the new cluster center using:

$$v_i = (1/c_i) \sum_{j=1}^{C_i} x_i \qquad 4.2$$

where, 'c_i' represents the number of data points in i^{th} cluster.

5) Recalculate the distance between each data point and new obtained cluster centers.

6) If no data point was reassigned then stop, otherwise repeat from step (3).

4.3 DECIDING THE NUMBER OF CLUSTERS

The number of clusters should match the data. An incorrect choice of the number of clusters will invalidate the whole process. An empirical way to find the best number of clusters is to try K-means clustering with different number of clusters and measure the resulting sum of squares.

K-Means clustering algorithm is applied to the power spectral density values estimated. The number of clusters is adjusted to get the optimum results. The distance metric used is Euclidean distance which is a standard one. The number cluster centriods are iteratively calculated to increase the similarity between points in the same clusters and to decrease the similarity between points in different clusters [19].

The various parameters are estimated for analyzing the performance efficiency of this algorithm is classifying epilepsy risk level. The performance parameters are namely Performance index, false alarm, Missed classification, Sensitivity, Specificity and Quality value. It is observed that the K-Means achieved false alarm rate of 0% and Sensitivity of 100%. It is desirable to get low false alarm rate, we have achieved that. The quality value of 22.37 is obtained with K-Means clustering which is greater than the value obtained with KNN Classifier.

The values obtained are tabulated below.

The advantages of K-Means clustering are given below.

- If variables are huge, then K-Means most of the times computationally faster than hierarchical clustering, if we keep k small.

- K-Means clustering produce tighter clusters the hierarchical clustering, especially if the clusters are globular.

The disadvantages of K-Means clustering are

- Difficult to predict K-value.
- With global cluster, it didn't work well.
- Different initial partitions can result in different final clusters.
- It does not work well with clusters (in the original data) of Different size and Different density.

Table 4.1 Performance parameters of K-Means clustering

Patient	PC (%)	MC (%)	FA (%)	PI (%)	Se (%)	Sp (%)	Av Detection (%)	Time delay (Second)	Qv
1	95.84	4.16	0	95.65	100	95.84	97.92	2.16	20.39
2	87.50	12.49	0	84.40	100	87.50	93.75	2.44	23.07
3	5.84	4.16	0	95.65	100	5.84	97.2	2.16	22.53
4	94.45	5.55	0	94.07	100	94.45	97.92	2.22	22.53
5	94.45	5.55	0	94.07	100	94.45	96.53	2.22	22.14
6	93.06	6.94	0	92.34	100	93.06	95.83	2.26	21.64
7	91.67	8.31	0	90.45	100	91.67	95.83	2.33	21.35

8	90.27	8.31	0	88.59	100	90.27	95.18	2.16	23.07
9	95.84	4.16	0	95.65	100	95.84	97.92	2.16	23.07
10	95.84	4.16	0	95.65	100	95.84	97.92	2.16	23.07
11	95.84	4.16	0	95.65	100	95.84	97.92	2.16	23.07
12	97.22	2.77	0	97.10	100	97.22	98.61	2.11	23.71
13	94.45	5.55	0	94.07	100	94.45	97.22	2.22	22.53
14	94.45	5.55	0	94.07	100	94.45	97.22	2.22	22.53
15	94.45	5.55	0	94.07	100	94.45	97.22	2.22	22.53
16	95.84	4.16	0	95.65	100	95.84	97.92	2.16	23.07
17	94.45	4.16	0	94.07	100	94.45	97.22	2.22	22.53
18	91.67	8.33	0	90.75	100	91.67	95.83	2.32	21.60
19	94.45	5.55	0	94.07	100	94.45	97.22	2.22	22.53
20	87.5	12.49	0	84.40	100	87.5	93.75	2.44	20.39
Average	93.75	6.10	0	93.02	100	93.75	96.87	2.24	22.37

CHAPTER 5
RESULT AND DISCUSSION

5.1 PERFORMANCE INDEX

In K- Means clustering the Perfect classification is about 93.75% which is very high when compared with KNN Classifier which is 83.33% only. The sensitivity and selectivity of K- Means clustering is also more when compared to the latter. K-Means clustering we have obtained a false alarm rate of 0% with K- Means clustering, whereas in KNN Classifier the value achieved is 9.50%. In K-Means clustering the Performance Index value is about 93.02% and 78.31%. Table 4.1 indicated the result details of KNN classifier and K-Means clustering.

The formulae for obtaining above specified values are given below

$$PI = \frac{PC - MC - FA}{PC} X\ 100$$

5.1

$$Sensitivity = \frac{PC}{PC + MC} X\ 100$$

5.2

$$Specificity = \frac{PC}{PC + MC} X\ 100$$

5.3

$$Accuracy = \frac{Se + Sp}{2} \times 100$$

5.4

where, PI - Performance index, PC - Perfect classification, MC - Missed Classification, FA - False alarm, Se - Sensitivity, Sp – Specificity.

5.2 QUALITY VALUE

In Order to compare different classifier we need a measure that reflects the overall quality of the classifier. Their quality is determined by three factors Classification rate, Classification delay, and False Alarm rate. The quality value Q_V is defined as

$$Q_V = \frac{C}{(R_{fa} + 0.2) * (T_{dly} * P_{dct} + 6 * P_{msd})} \quad 5.5$$

Where, C is the scaling constant

R_{fa} is the number of false alarm per set

T_{dly} is the average delay of the onset classification in seconds

P_{dct} is the percentage of perfect classification and

P_{msd} is the percentage of perfect risk level missed

A constant C is empirically set to 10 because this scale is the value of Q_V to an easy reading range. The higher value of Q_V, the better the classifier among the different classifier, the classifier with the highest Q_V should be the best.

The table 4.1 specifies the average performance of the KNN Classifier and K-Means Clustering with the PSD dimensionality reduction. The results show that the performance of the K-Means Clustering with PSD dimensionality reduction has best performance in terms of Missed Classification, False Alarm, Sensitivity, Specificity, Performance Index and Average Detection.

Table 5.1 Performance Comparisons of KNN Classifier and K-Means Clustering

Performance Parameters	KNN Classifier	K-Means Clustering
Perfect Classification (%)	83.3325	93.7540
Missed Classification (%)	7.1485	6.1077
False Alarm (%)	9.5029	0
Performance Index (%)	78.3193	93.0242
Sensitivity (%)	0.4877	100
Specificity (%)	92.8475	93.8771
Average Detection (%)	87.0948	96.8771
Time delay (second)	2.107	2.2427
Quality value	18.0206	22.3728

These models are evaluated in terms of its receiver operating characteristics (ROC) curve for test data sets. This enables the user to evaluate a model in terms of the trade-off between sensitivity and specificity. ROC matrices are used to show how changing detection threshold affects detection versus false alarms. The percentage of detections classified correctly is plotted against the percentage of non-detections in correctly classified as detections (i.e. false alarms) as a function of the detection threshold. ROC is the best way to evaluate a detector.

The comparison plots are given below.

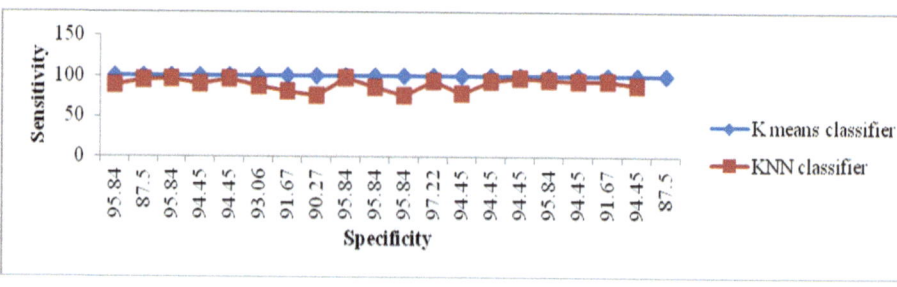

Figure 5.1 Sensitivity and Specificity measures of KNN classifier and K- means clustering

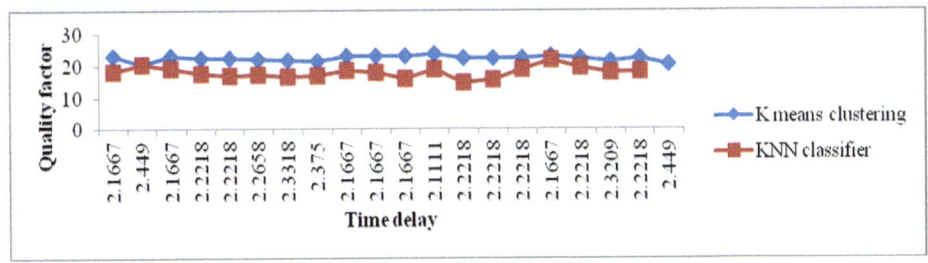

Figure 5.2 Quality factor and Time delay measures of KNN classifier and K- means clustering

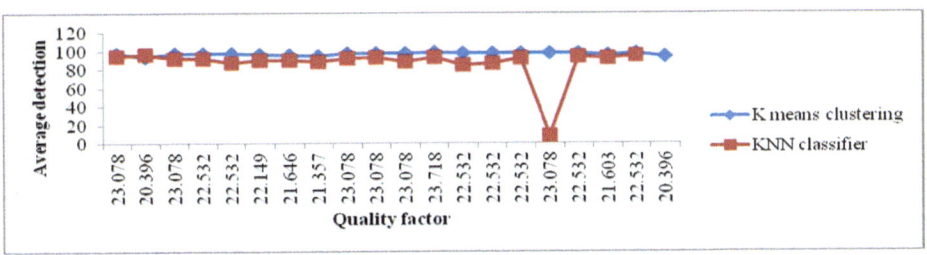

Figure 5.3 Average detection and Quality factor measures of KNN classifier and K- means clustering

It has been found that, KNN Classifier got Quality value of 18.02, and K- Means achieved the value of 22.37. Hence K-Means Classifier outperforms KNN classifier in Epilepsy risk level classification.

CHAPTER 6
CONCLUSION AND FUTURE EXPANSIONS

This Project investigates the performance of KNN Classifier and K-Means Clustering in optimizing the epilepsy risk level of epileptic patients from EEG signals. The parameters derived from the EEG signal are stored as data sets. The objective was to obtain

- High rate of classification
- High Performance Index
- High Quality Value
- Low False Alarm
- Low Missed Classification
- A short delay from onset

Though it is impossible to obtain a perfect performance in all these conditions, some compromises have been made. As a high false alarm rate ruins the effectiveness of the system, a low false-alarm rate is most important. The Power spectral density is estimated for feature analysis. Dimensionality reduction is carried out for these spectral values. Further KNN classifier and K-Means Clustering are applied to dimensionality reduced feature values. K-Means clustering algorithm achieved high performance Index than K-Means Clustering.

The number of cases from the present twenty patients has to be increased for better testing of the system. From this method we can infer the occurrence of High-risk level frequency and the possible medication to the patients. Also optimizing each region's data separately can solve the focal epilepsy problem.

The further studies are in the direction of a comparison of SVM method between KNN Classifier and K-Means Clustering optimization models.

REFERENCES

1. Ye Yaun, "Detection of epileptic seizure based on EEG signals", Image and Signal Processing (CSIP), 2010 3rd International Congress on, vol. 9, pp. 4209 – 4211, October 2010.
2. Sunil Kumar Prabhakar, Harikumar Rajaguru, "Application of Linear Graph Embedding as a Dimensionality Reduction Technique and Sparse Representation Classifier as a Post Classifier for the Classification of Epilepsy Risk Levels from EEG Signals", Proceedings of the International Conference on Graphic and Image Processing (ICGIP), October 23-25, 2015, Singapore.
3. Sunil Kumar Prabhakar, Harikumar Rajaguru, " An Efficient Classification of Epilepsy Risk Levels from EEG Signals Using Hard Thresholding Computation Applied to Code Converters", Proceedings of 2015 IEEE Student Symposium in Biomedical Engineering and Sciences (ISSBES), November 4, 2015, Universiti Teknologi Mara, Malaysia
4. Sunil Kumar Prabhakar, Harikumar Rajaguru, "Performance Comparison of Fuzzy Mutual Information as Dimensionality Reduction Techniques and SRC, SVD and Approximate Entropy as Post Classifiers for the Classification of Epilepsy Risk Levels from EEG Signals", Proceedings of 2015 IEEE Student Symposium in Biomedical Engineering and Sciences (ISSBES), November 4,2015, Universiti Teknologi Mara, Malaysia.
5. Sunil Kumar Prabhakar, Harikumar Rajaguru, "A Different Approach to Epilepsy Risk Level Classification Utilizing Various Distance Measures as Post Classifiers", Proceedings of the 8th Biomedical Engineering International Conference (BMEiCON), November 25-27, 2015, Pattaya, Thailand.
6. Sunil Kumar Prabhakar, Harikumar Rajaguru, "A Novel Combination of Code Converters and Sparse Representation Classifier for an Efficient Epilepsy Risk Level Classification", Proceedings of the 8th Biomedical Engineering International Conference (BMEiCON), November 25-27,2015, Pattaya, Thailand.
7. Sunil Kumar Prabhakar, Harikumar Rajaguru, "Morphological Operator Based Feature Extraction Technique along with Suitable Post Classifiers for Epilepsy Risk Level", Proceedings of the International Conference on Intelligent Informatics and BioMedical Sciences (ICIIBMS), November 28-30, Okinawa, Japan.
8. Sunil Kumar Prabhakar, Harikumar Rajaguru, "PCA and K-means Clustering for Classification of Epilepsy Risk Levels from EEG Signals – A Comparative Study Between Them", Proceedings of the International Conference on Intelligent Informatics and BioMedical Sciences (ICIIBMS), November 28-30, Okinawa, Japan.
9. Sunil Kumar Prabhakar, Harikumar Rajaguru, "GMM Better than SRC for Classifying Epilepsy Risk Levels from EEG Signals", Proceedings of the International Conference on Intelligent Informatics and BioMedical Sciences (ICIIBMS), November 28-30, Okinawa, Japan.

10. Sunil Kumar Prabhakar, Harikumar Rajaguru, " Cascaded Feed Forward Neural Networks and Generalized Regression for Epilepsy Risk Level Classification – A Study" Proceedings of the 3rd MEC International Conference on Big Data and Smart City (ICBDSC), March 15-16, 2016, Muscat, Oman
11. M. Heers, S. Rampp, M. Kaltenhauser, E. Pauli, C.Rauch, M.T. Dolken, H.Stefn, "Detection of Epileptic Spikes by Magnetoencephalography and Electroencephalography after sleep deprivation", Seizure, vol. 19, Issue. 7, pp. 397- 403, September 2010.
12. A.S.Gevins, "Analysis of the electromagnetic signals of the human brain: Milestones, Obstacles, and Goals, "IEEE Trans. Biomed Eng, vol. 31, pp. 833 850, 1984.
13. Arunasakthi. K, KamatchiPriya. L, "A Review On Linear and Non – Liner dimensionality Reduction techniques, Machine Learning an Applications: An International Journal(MLAIJ) vol. 1, No. 1, September 2014.
14. Tian Siva Tian, Gareth M.James, "Interpretable Dimesion Reduction for Classifiying Function Data",omputional Statistics and Data Analysis. vol. 57, Issue. 1, pp. 282 – 296, January 2013.
15. N.Faud, M.N.Taib, R.Jailani, M.E.Marwan, " Brainwave classification for Brain Balancing Index (BBI) via 3D EEG Model using KNN technique, International Journal of Computer, Electrical, Automation, Control and Information Engineering, vol. 8, No. 8, 2014.
16. R. Shantha Selva Kumari and J.Prabin Jose, " Seizure detection in EEG using Biorthogonal wavelet and fuzzy KNN classifier", Elixir Human Physio. vol. 42, pp. 5766 – 5770. 2011.
17. Belur V. Dasarathy, " Nearest Neighbor (NN) Norms: NN Pattern Classification Techniques", Mc Graw-Hill Computer Science Series, IEEE Computer Society Press, Las Alamitos, California, pp.217-224,1991.
18. N.Suguna, Dr.K. Thanushkodi, " An Improved K- Nearest Neighbor Classification Using Genetic Algorithm", IJCSI International Journal of Computer Science Issues, vol. &, Issue. 4, No. 2, July 2007.
19. Jennifer G.Dy , Carla E. Brodley, " Feature Selection for Unsupervised lerning", Journal of Machine Learning Research 5 (2005), pp. 845 – 889.
20. Dr.R.Harikumar, T.Vijayakumar, M.G. Sreejith, "Performance Analysis of SVD and K-Means Clustering for Optimization of Fuzzy Outputs In Classification of Epilepsy Risk Level from EEG Signals", Recent Advances In Intelligent Computational Systems (RAICS), 2011 IEEE, pp. 718 – 723.
21. K P Adlassnig (March 1986), 'Fuzzy Set Theory in Medical Diagnosis', IEEE Transactions on Systems Man Cybernetics, 16: 260-265.
22. Arthur C Gayton (1996), 'Text Book of Medical Physiology', Prism Books Pvt. Ltd., Bangalore, 9th Edition.

23. Celement.C etal (April 2003), 'A Comparison of Algorithms for Detection of Spikes in the Electroencephalogram',IEEE Transaction on Bio Medical Engineering, 50 (4): 521-26.
24. Haoqu and Jean Gotman (February 1997), 'A Patient Specific Algorithm for Detection Onset in Long-term EEG Monitoring Possible use as Warning Device', IEEE Transactions on Biomedical Engineering, 44(2): 115-122.
25. R.Harikumar, Dr. (Mrs). R.Sukanesh and P.A. Bharathi (May 2005), 'Genetic Algorithm Optimization of Fuzzy outputs for Classification of Epilepsy Risk Levels from EEG signals', I.E. India Journal of Interdisciplinary panels, Vol.86, no.1, pp.12-20.
26. S.Haykin (1999), 'Neural Networks a Comprehensive Foundation', Prentice- Hall Inc. 2nd Ed..
27. Joel.J etal (April 2004), 'Detection of Seizure Precursors from Depth EEG Using a Sign Periodogram Transform', IEEE Transactions on Bio Medical Engineering, 51 (4):449-458.
28. Leon D.Iasemidis (May 2003), 'Adaptive Epileptic Seizure Prediction System', IEEE Transactions on Biomedical Engineering, 50(5): 616-627.
29. Mark van Gils (November/December 1997), 'Signal Processing in Prolonged EEG Recordings During Intensive Care', IEEE EMB Magazine ,16(6): 56-63.
30. Melanie Mitchell (1997), 'An Introduction to Genetic Algorithms', A Bradford Book MIT Press.
31. Mu-chun Su, Chien –Hsing Chou (June 2001), 'A Modified Version of the k-means Clustering Algorithm with a Distance Based on Cluster Symmetry', IEEE Transactions on Pattern Analysis and Machine Intelligence, 23 (6): 674-680.
32. Rangaraj M. Rangayyan (2002), 'Bio- Medical Signal Analysis: A Case Study Approach', IEEE Press-John Wiley &sons Inc New York
33. Richard O. Duda, David G. Stroke, Peter E. Hart (2003), 'Pattern Classification, Second Edition, Wiley-Interscience Publication, John Wiley and Sons, Inc.
34. J.Seunghan Park et al (August 1990), 'TDAT Domain Analysis Tool for EEG Analysis', IEEE Transactions on Biomedical Engineering, 37(8): 803-811.
35. Sathish Kumar (2004), 'Neural Networks, A Classroom Approach', McGraw-Hill New York.
36. V.Vapnik (1998), 'Statistical Learning Theory', Wiely Chichester, GB